j363.7 S596c
Climate action :what happened
and what we can do /
Simon, Seymour,

MAY 2 5 2021

WITHDRAWN

D1120111

CLIMATE ACTION

WHAT HAPPENED AND WHAT WE CAN DO

SEYMOUR SIMON

HARPER

An Imprint of HarperCollinsPublishers

To Liz, my wife and partner
in conceiving and researching this book.

Climate Action: What Happened and What We Can Do   Copyright © 2021 by Seymour Simon   All rights reserved.   Manufactured in Italy.
No part of this book may be used or reproduced in any manner whatsoever without written permission except in the case of brief quotations embodied in critical articles and reviews. For information address HarperCollins Children's Books, a division of HarperCollins Publishers, 195 Broadway, New York, NY 10007.
www.harpercollinschildrens.com

ISBN 978-0-06-294331-6 (trade bdg.) — ISBN 978-0-06-294330-9 (pbk.)

Photo Research by Liz Nealon   20 21 22 23 24   RTLO   10 9 8 7 6 5 4 3 2 1   ❖   First Edition

AUTHOR'S NOTE

In 1972, in the early days of the environmental movement, I wrote my first two books about protecting Earth. They were called *Science Projects in Pollution* and *Science Projects in Ecology*. At that time, helping our planet Earth meant cleaning up our air, soil, and water. We were only beginning to understand how our own activities—and many of the industries created since the Industrial Revolution—were damaging the environment.

Almost forty years later, in 2010, I was invited to speak at a rally on the National Mall in Washington, DC, for Earth Day's fortieth birthday. I had written a new book called *Global Warming*, and in my speech, I said that in 2007, a report by 2,500 scientists from 130 countries concluded that humans are responsible for much of the current warming. But before I finished, a man at the front of the crowd shouted, "LIAR!" Incredible! How surprised I was to think that someone didn't believe the facts and the science behind global warming.

I don't think that would happen today, because most people—especially the young, inspiring ones you'll meet in this book—

now agree that there is a clear danger to our environment and our planet. As serious as the situation has become, it is not too late to take action together. I am inspired by the courage of the young activists who have captured the attention of the whole world. Once we educate ourselves about the science, we can all find ways to do our part and persuade others to take care of Earth, our home planet.

"My name is Greta Thunberg. I am fifteen years old. I am from Sweden. I speak on behalf of Climate Justice Now. Many people say that Sweden is just a small country and it doesn't matter what we do. But I've learned you are never too small to make a difference. And if a few children can get headlines all over the world just by not going to school, then imagine what we could all do together if we really wanted to."

GRETA'S CHALLENGE TO THE WORLD

Greta Thunberg spoke those words to the assembled delegates at the UN Climate Change Conference in 2018. She is a young activist who at age fifteen began protesting about the need for immediate action to combat climate change. Her speeches and actions were the beginning of the global School Strike for Climate movement, which led to a worldwide climate strike in September 2019. Organizers estimated that over 7 million people from every continent participated. The purpose of the strike was to urge responsible adults to take action to stop climate change.

Greta also addressed the United Nations Climate Summit in 2019, traveling from Europe in a solar-powered boat so as not to increase her carbon footprint by flying in an airplane. Her speech to world leaders assembled at the UN was searing and powerful. Here is some of what she said:

"My message is that we'll be watching you. This is all wrong. I shouldn't be up here. I should be back in school on the other side of the ocean. Yet you all come to us young people for hope. How dare you. . . . For more than thirty years, the science has been crystal clear. How dare you continue to look away and come here saying that you're doing enough when the politics and solutions needed are still nowhere in sight. . . . We will not let you get away with this. Right here, right now is where we draw the line. The world is waking up. And change is coming, whether you like it or not."

GLOBAL WARMING OR CLIMATE CHANGE?

Whether you call what's happening *global warming* or *climate change*, it's still the biggest challenge humanity has ever faced. Global warming causes climate change, so the two terms are easy to confuse. *Global warming* is used to describe the long-term trend of higher average temperatures around the world. *Climate change* is an alteration in normal climate patterns usually due to increasing levels of carbon dioxide in the air.

Practically all scientists agree that climate change is happening and planet Earth is getting warmer every year. According to weather scientists, the years from 2015 to 2019 were the hottest five years on record. In 2019, temperature records were shattered around the world.

39262934
MOM9460172
9780062943316
l 42o5oooooooo
[1]Custom:ROUTE TO COVER-UP - CHECK
NOTE FOR JUV/YA SPINE TAPE
SPINE TAPE PBK ONLY

j363.7 S596c

INTENTIONALLY
EMPTY

10900061697280

SPINE TAPE PBK ONLY

MPC

MESA COUNTY PUBLIC
LIBRARY DISTRICT
P.O. BOX 20000/530 Grand Ave.
GRAND JUNCTION, CO 81502

IS CLIMATE CHANGE REAL?

ALESSANDRO DAL BON was one of the New York City organizers of the 2019 Climate Strike, even though he was not yet old enough to vote.

Like Alessandro, you may have heard some grown-ups claiming that climate change doesn't exist. But the fact is that a vast, overwhelming majority of scientists believe that climate change is real and we have to do something about it. You might think you have to be an adult to make an impact on climate change, but you don't.

"We need to stop asking 'Do you believe in climate change?' and start asking 'How do we stop climate change?' The climate crisis is not a belief. It is a fact, and it has been proven by the science. So let's start acting like it is."

In recent times, climate change has been linked to politics, but climate change is not a political issue. It is a human issue that affects all of us. So it's more important than ever to use science-based evidence to sort fact from fiction.

In this book, we'll explore some of the most serious effects of climate change, learn how humans have contributed to many of these problems, and explore what is being done to try to address the situation. We will also talk about what you can do in your own life to help fight climate change.

WEATHER VS. CLIMATE

There is a difference between weather and climate. Weather happens over a short period of time, such as days or even hours. A weather forecast may say that it will rain or snow tomorrow afternoon. Climate is the long-term average of weather patterns over years, decades, and centuries. A climate may be tropical and dry, or cold and snowy. Here is an easy way to remember: climate is what you expect (a hot summer or a cold winter), and weather is what you actually get (a day of rain or a snowstorm).

7

WHAT IS THE GREENHOUSE EFFECT?

Climate change and global warming are happening because of the greenhouse effect. A greenhouse is a building with glass walls and a glass roof used to grow plants and flowers. In the daytime, the sun's rays go through the glass and warm the air and plants inside. In the nighttime, the glass walls keep the warm air from escaping, and it remains warm inside, even during the winter.

HOW IS EARTH LIKE A GREENHOUSE?

Our planet Earth is not a greenhouse, but certain gases in the atmosphere act like the glass of a greenhouse. Sunlight passes through the atmosphere and warms the ground. Some of the heat bounces back and is radiated into space. But much of the heat remains trapped by GHGs.

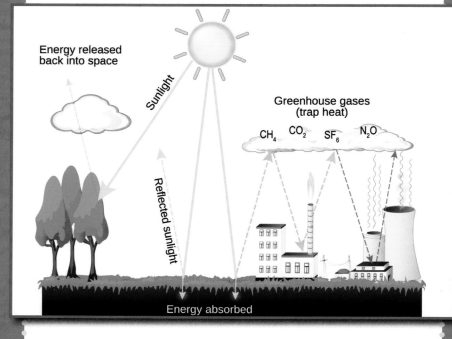

Energy released back into space

Sunlight

Greenhouse gases (trap heat)

CH_4 CO_2 SF_6 N_2O

Reflected sunlight

Energy absorbed

GREENHOUSE GASES:

☀ are gases in Earth's atmosphere that trap heat

☀ are also known as GHGs

☀ include water vapor, carbon dioxide, methane, chlorofluorocarbons

☀ are good because they keep the planet warm enough for life to exist

☀ can be bad because too many GHGs trap heat and cause global warming

☀ have been released into the atmosphere in large amounts since the late 1800s

WHY ARE GREENHOUSE GASES INCREASING?

Greenhouse gases started to build up at the beginning of the industrial age over 150 years ago, when the rise of factories and heavy industries, the cutting down and destruction of forests, and the introduction of large-scale farming spewed forth more greenhouse gases than ever before. We have also greatly increased the amount of carbon dioxide in the air by burning fossil fuels such as coal, gasoline, and oil.

XIYE (SHE-yuh) **BASTIDA PATRICK** knows firsthand about the effects of too many GHGs. She grew up in the small town of San Pedro Tultepec, outside of Mexico City. When her community suffered through years of drought followed by severe flooding that destroyed crops and businesses, Xiye's family immigrated to New York City. Shortly after they arrived, New York's coastal communities were devastated by flooding from Hurricane Sandy. Because of these experiences, Xiye has become a leader in the youth climate crisis movement.

"My generation is called Gen Z. . . . Z is the last letter of the alphabet. It symbolizes the end of something. Does that mean you want us to be the last generation? I don't think it does. So we are reframing that and saying: We are going to be the last generation that is dependent on fossil fuels."

WHAT HAPPENS WHEN WE HAVE CLIMATE CHANGE?

HEAT WAVES

According to the *Guinness World Records* book, the world's largest water balloon fight took place on August 27, 2011. It happened at the University of Kentucky, and the nearly 9,000 people who took part used 175,141 water balloons!

ALL-TIME RECORD HIGH
ANCHORAGE, AK

NEW RECORD
90°
JULY 4, 2019

OLD RECORD
85°
JUNE 14, 1969

Old Record High for July 4:
77° set in 1999.

AccuWeather

What ever possessed all those students to stage such a massive water balloon tossing event? We'll never know for sure, but it probably had something to do with the fact that North America was in the midst of the deadly 2011 heat wave, the hottest summer in seventy-five years.

A heat wave is a stretch of two or more days of abnormally and uncomfortably hot and unusually humid conditions. The temperatures must be hotter than the usual temperatures of the area. For example, a few days of 95°F temperatures in summer in Alaska or Maine would be a heat wave, but a couple of 95°F summer days in Death Valley, California, would be unremarkable.

In many places around the world in recent years, people have experienced temperatures never recorded before. In Anchorage, Alaska, temperatures in July 2019 soared to over 90°F, the highest ever seen in that city. In Europe in the summer of 2019, the daily temperatures were the highest ever seen there.

HOW MUCH WARMER HAS EARTH BECOME?

According to the United States National Oceanic and Atmospheric Administration (NOAA), Earth has had above-average temperatures for more than forty straight years (since 1977). Not only that, but the past five years have been the warmest since record keeping began in the late nineteenth century.

This chart shows the average monthly carbon dioxide measurements at the Mauna Loa Observatory in Hawaii since 1958. The increase in carbon dioxide usually results in an increase in temperatures.

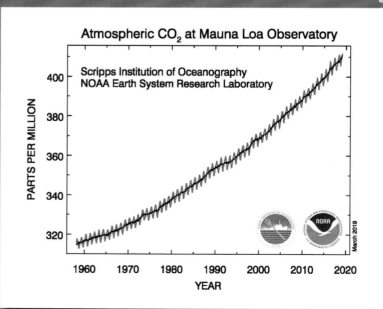

REDUCING GREENHOUSE GASES: WHAT WE CAN DO

Develop renewable energy sources

One of the main ways to reduce the amount of greenhouse gases is to make electricity without using fossil fuels such as oil and coal. Electricity can be made with solar power, wind power, nuclear power, geothermal power (underground heat), and even tidal power (using the energy of daily ocean tides).

Drive electric vehicles (EVs)

Gasoline-powered cars and trucks produce about one-fifth of greenhouse gases. Replacing gas- and diesel-powered cars with electric and hybrid engines can reduce carbon-dioxide emissions. To make a practical electric car means developing and making batteries that can hold a charge for long driving times.

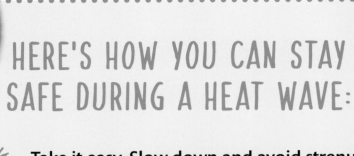

HERE'S HOW YOU CAN STAY SAFE DURING A HEAT WAVE:

☀ Take it easy. Slow down and avoid strenuous games or exercise.

☀ Dress for the weather. Wear lightweight, light-colored clothing.

☀ Drink lots of water. Keep hydrated as much as possible.

☀ Find an air-conditioned place to hang out. Don't have an air conditioner at home? Find a friend with one or go to a public place (like a public library) that has one.

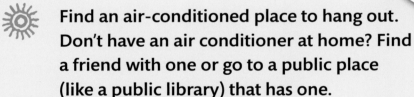

☀ Avoid too much sun. Sunburns make it harder for your body to cool down.

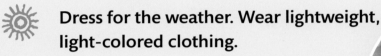

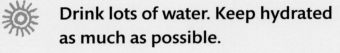

☀ Never stay in a parked car with the windows closed or leave your sibling or pet there.

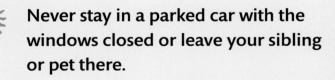

14

THE DEVASTATING IMPACT OF DROUGHTS

KYNAN TEGAR is from Sungai Utik, Indonesia, a country in Southeast Asia. The region has suffered increasingly long drought seasons, in recent years going months without rain.

As Kynan experienced in Indonesia, droughts are long periods of unusually dry weather that cause crop damage, water shortage, and forest fires. Droughts threaten food supplies, water availability, and the health of all living things—plants, animals, and humans.

"In my community we have been really feeling the effects of climate change. The drought season is getting longer and longer, which is creating a lot more forest fires. So, this climate crisis threatens our trees, our land, and our culture. Our culture is intrinsically linked to our forests. Without the forest, we have no culture."

Farmers carry buckets of water to irrigate their fields in Boyolali, Central Java, Indonesia.

TOO MUCH HOT AIR

Droughts happen because as average temperatures rise, the warmer air can absorb and hold more moisture, which means there is less rain. Hotter air also causes more water to evaporate from lakes and rivers, so there is less water for both humans and plants. Plants normally help to keep moisture in the soil, but if they die from lack of water, conditions become even drier.

As more extreme climate patterns emerge, there is a shift away from light, average rainstorms that last less than an hour to periods of short, very heavy rains that last a few hours and run off quickly because the drier soils are less able to absorb the water. This leads to an increased possibility of flooding. In some places, floods can develop in just a few minutes.

2050: THE LOOMING MEGADROUGHT

NOAA research shows that humans have been driving global patterns of drought for nearly a century as greenhouse-gas emissions have steadily increased. Scientists at Cornell University predict that there is at least an 80 percent chance that by 2050 the American Southwest and Great Plains will experience a "megadrought" that will last for fifty years.

Average US Temps 2050–2059

Average maximum temperature (°F)

10   60   110

Climate.gov
Data: LOCA

CAN WE AVOID DROUGHTS?

Scientists are working now to develop new methods for reducing the effects of drought, including inexpensive desalinating of (removing salt from) ocean water so that it is usable to support life and harvesting water from air. There are many solutions being developed, some already in use. These include:

PLANTING MORE TREES: This is the best way to help soil retain water, as the deep roots of trees store moisture.

RAINWATER HARVESTING: Both homes and communities can capture and store the water that falls when it rains and use it later when conditions are dry.

RECYCLING ORGANIC WASTE: Add organic materials like lawn clippings, leaves, cornstalks, manure, and clean food waste (such as vegetable peels) to the ground to help dry soil retain water.

WILDFIRES: OUT OF CONTROL

ALEXANDRIA VILLASEÑOR is familiar to many as the girl who for nearly a year sat on a bench outside United Nations headquarters every Friday, holding signs about the need for climate action. The issue is personal for Alexandria. When she was a young teenager, her family went on a camping trip in Northern California and were enveloped by a smoke cloud from the historic 2018 Camp Fire. Alexandria, who has asthma, became very ill and frightened. Once she started doing research about the causes of these massive wildfires, she realized that they are one of the big effects of climate change.

A wildfire (also known as a "bushfire" in some places) is an out-of-control fire that burns in a natural area such as a forest, grassland, or prairie, wiping out vegetation and animal life. Wildfires most often occur in areas that are in drought, where warmer air, dead vegetation, and decreased rainfall all combine to create a landscape that is ripe for wildfire.

DESTRUCTIVE WILDFIRES IN THE UNITED STATES

The frequency of wildfires in the western United States has increased by 400 percent since 1970. This is a direct consequence of global warming.

The 2018 wildfire season in California was the deadliest and most destructive ever in the US, burning nearly two million acres, destroying homes and vehicles, and killing 103 people.

WILDFIRES AROUND THE WORLD

In the following years, wildfires were burning all over the world, even in cold places like Alaska and Siberia, where fire is not usually a major threat.

For years, scientists had warned that climate change would increase the risk and severity of fires in Australia. They were right. The raging Australian bushfires in the summer of 2019 killed a billion animals and caused human deaths and widespread destruction.

Volunteer rescuing koala from Australian bushfires.

HUMANS CAUSE WILDFIRES

More than 85 percent of wildfires are caused by humans. Since the land and vegetation in drought-plagued areas are so dry, even carelessly discarding a lighted cigarette butt can start a fire. Major wildfires can also be ignited by trees and branches touching electrical power lines. This was the cause of California's deadliest wildfire, the 2018 Camp Fire, in which eighty-five people died and nearly 19,000 homes and other buildings were destroyed.

OTHER CAUSES OF WILDFIRES:

- Unattended campfires
- Embers from fireworks falling on dry brush
- Burning of trash or other waste rather than disposing of it safely
- Unfortunately, sometimes humans set wildfires deliberately.

FIREFIGHTING DRONES

Engineer and former Myth-Buster **JAMIE HYNEMAN** is working on a new kind of driverless vehicle: a firefighting tank called the Sentry, fitted with massive water containers and piloted by remote control. Hyneman hopes it will be helpful in fighting wildfires since it will be able to roll directly into flames too intense for a human firefighter and crush smoldering debris under its heavy treads.

Hyneman isn't the only one working on engineering a solution to wildfires. Engineers and scientists are using existing technologies in new ways and inventing new equipment and methods to prevent, detect, and fight wildfires. California is developing a system called FUEGO (Fire Urgency Estimator in Geosynchronous Orbit). *Fuego* is the Spanish word for fire, and FUEGO uses a satellite and drones to spot early-stage wildfires, tracks the fire's progress, and alerts firefighters on the ground before it burns out of control.

FIREFIGHTING ROBOTS AND SMOKE JUMPERS

Scientists and engineers are also developing firefighting robots that can go into areas that are too hot and too dangerous for humans to penetrate. And organizations around the world are using virtual reality (VR) simulations to train firefighters and smoke jumpers (firefighters who parachute into remote areas to combat wildfires) in safe conditions.

21

PERMANENT DESTRUCTION

DEFORESTATION

ARTEMISA XAKRIABÁ was just seven years old when she and other students from her indigenous tribe planted trees, helping to reforest riverside areas near their traditional lands in Brazil. Now a young adult, Artemisa traveled to New York to join the 2019 Climate Strike.

"We are the main guardians of the forest. We, indigenous people, depend on nature . . . that is what keeps us alive. . . . If it goes on like this, twenty years from now my homeland will become a desert."

Scientists estimate that there are 3 trillion trees on Earth, an incredibly large number. There are more trees on our planet than there are stars in the Milky Way galaxy.

Scientists often describe trees as "Earth's lungs" because they absorb and store carbon dioxide and other harmful greenhouse gases and release oxygen into the air.

Deforestation is the permanent destruction of forests to make land available for other uses, such as agriculture or building homes. Fewer forests and trees mean less carbon dioxide is being absorbed and turned into oxygen. When trees die, they release harmful greenhouse gases that had been stored in their trunks for years. Not only do animals lose their habitats when we lose trees, we lose the single greatest weapon we have against the greenhouse effect.

Recently cut and burned area of the Amazon rainforest turned into a cattle ranch.

23

SACRIFICING TREES FOR FOOD

Much of the fruit and grain we buy from tropical countries is grown in areas that were once rainforest. The forests are cut down to make way for vast plantations where products such as bananas, sugarcane, tea, and coffee are grown. The wet soil from the rainforest supports agriculture for a few years. But once the rainforest trees are gone, the rich soil washes away and is no longer suitable for farming.

THE PRICE OF EATING MEAT

As Brazil is the world's largest exporter of beef, most fires in the Amazon are set because ranchers want to clear the forest to make room for grazing cattle. Cattle require large amounts of land for grazing. Scientists estimate that for each pound of beef produced, 200 square feet of rainforest are cut down. That means that the more we eat beef in our diets, the more trees are killed.

In 2019, more than 36,000 separate fires devastated the Amazon rainforest. Deforestation in the Amazon increased by 80 percent compared to the previous year, and there was so much smoke in the air that Brazil's largest city, São Paolo, experienced days when it was as dark as night in the middle of the afternoon.

THE PARIS CLIMATE AGREEMENT

The main goal of the 2015 Paris Climate Agreement is to limit the global temperature increase to 1.5°C (2.7°F). The IPCC report looked at land-use practices that have impacted climate change and found that cutting down trees and forests, meat-based agriculture, and other human activities threaten the world's ability to make that happen. If more people around the world shift their eating habits, it could boost our ability to fight climate change.

WHAT WE CAN DO

The choices we make about what we eat and how we live affect climate change.

PLANT TREES: Every animal on Earth, including human beings, requires oxygen. Trees produce oxygen and take in carbon dioxide, so protecting our forests and planting more trees is critical to fighting climate change.

Once a tree is ten years old, it is at the most productive stage of carbon dioxide absorption and can absorb nearly fifty pounds of carbon dioxide a year. So planting a tree helps. Even a single tree.

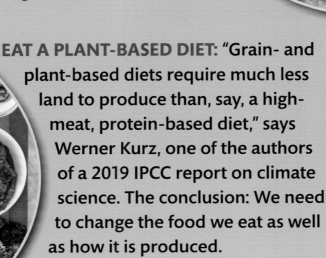

EAT A PLANT-BASED DIET: "Grain- and plant-based diets require much less land to produce than, say, a high-meat, protein-based diet," says Werner Kurz, one of the authors of a 2019 IPCC report on climate science. The conclusion: We need to change the food we eat as well as how it is produced.

WHAT CAN KIDS DO?

> "We are not too young to understand the science. Just because we can't vote doesn't mean we don't deserve a seat at the table, especially when the topic of discussion is our futures."

Teen climate activist ISHA TOBIS CLARKE, a high school student from Oakland, California, met with her senator, California Democrat Dianne Feinstein, to ask her to vote for the Green New Deal.

The viral video of Isha and other teen activists trying to discuss climate change with Senator Feinstein has been viewed over ten million times on Twitter.

> "To have to fight for the security of our future is a huge shame on all political leaders, present and past. What we're asking for is courage, from our political leaders and from these judges, because we're asking for an end to the system as usual."

In 2015, KELSEY JULIANA, along with twenty-one other kids and teens, filed a lawsuit, *Juliana v. United States*, which claimed that the US government knowingly failed to protect them against climate change.

Indigenous Canadian teenager **AUTUMN PELTIER** has been appointed the chief water commissioner for the Anishinabek Nation. She is known as the "Water Warrior" and spoke at the UN in 2019 about the importance of conserving and protecting our water supply.

"Nothing can live without water. If we don't act now, there will come a time when we will be fighting for those last barrels of water. Once that's gone, we can't eat or drink money or oil. Then what will you do?"

EVERY JOURNEY BEGINS WITH A SINGLE STEP

SPREAD THE WORD

The main point of talking about climate change with your parents and other adults is to get them to pay attention to the problem. Make it clear that fighting climate change is something that you're excited about.

CONTACT YOUR CONGRESSPERSON AND YOUR TWO SENATORS

Ask someone in your family, your teacher, or your librarian to help you find the address or phone number for each representative of yours. If you decide to call on the phone, write down what you want to say in advance so you will be able to make your point clearly.

SUPPORT GREEN CANDIDATES

Explain to your family that supporting political candidates who care about climate change is important for the future world that you will live in.

CLIMATE ACTION CHECKLIST

 TURN THINGS OFF

Shut off the lights when you leave a room; turn off your computer and TV if you are not using them. By using less electricity, you lessen the amount of carbon dioxide released by power plants.

 CLOSE DOORS

Heating or air-conditioning uses much less electricity if you keep the doors to the outside closed. Even keeping the fridge open while you decide what to eat wastes energy.

 BE A WATER SAVER

Turn off the faucet while brushing your teeth. Take shorter showers. Only run the washing machine and dishwasher when there is a full load.

 DON'T WASTE FOOD

About one-third of all the food produced in the world is thrown away, which means we are wasting the energy and water that it took to produce it. Worse yet, when discarded food decays in a landfill, it produces methane, a greenhouse gas.

 WEAR LAYERS

Put on a sweater instead of turning up the heat. Setting the thermostat just two degrees lower in the winter and two degrees higher in the summer saves about two thousand pounds of carbon dioxide per family per year.

 RECYCLE AND REUSE

You can recycle aluminum cans, glass, and things made of paper, such as cardboard, newspapers, and magazines. If your local government has set up recycling centers, use them! If they haven't, perhaps your school could urge them to get a program started.

EXTREME WEATHER

Early one August morning in 2017, ten-year-old **ELIRAN**, fourteen-year-old **RON**, and fifteen-year-old **SHAKED PLUSHNICK-MASTI** were roused out of sleep by their parents. Their house was flooding after days of torrential rain dumped by Hurricane Harvey, a slow-moving weather system that stalled over southern Texas. The family called 911, but the fire department couldn't say when they would arrive. The family gathered a few belongings in plastic bags and waited as the water rose above their knees, furniture began to topple over and float, and the street in front of their house became a roaring river. Finally, at 2 p.m. a boat arrived to ferry them to a safe shelter.

When Hurricane Harvey stalled over the city of Houston, it dumped more than 50 inches of rainfall (an entire year's worth of rain) in some places, making it the wettest Atlantic hurricane ever measured. The weight of the water was so great that it temporarily pushed down the city by almost an inch.

Heavy storms and flooding are happening more frequently because weather patterns all around the world are getting more extreme all the time. This is largely due to higher air temperatures, warmer ocean waters, and melting ice at the poles. A recent scientific study estimated that nearly 20 percent of heavier rains are the result of global warming. Even more important, the study found that the number of extreme weather events will increase as average world temperatures increase.

Floods are often caused by heavy storms. Cities are made of concrete, asphalt, steel, and other nonporous materials. These cities have miles and miles of buildings and flat, paved roads and walkways, leaving water nowhere to go but up. The water will eventually evaporate back up into the atmosphere—but that takes a long time.

Shaked Plushnick-Masti (right) and his friend Ethan Towber put Shaked's ruined belongings into garbage bags after Hurricane Harvey.

Green building in Paris, France.

WATER, WATER, EVERYWHERE

There is another way to think about dealing with heavy rains: make space for the water. City planners believe that we should start to redesign our coastal communities to work *with* floodwater, capturing it for future consumption even as we prevent it from overtaking our buildings, roadways, and homes.

Ways to do that include "renaturing," which means converting paved land to parks and planting grass on rooftops so that all the new green areas can absorb water like a sponge. We then will use these green spaces to capture rainwater and store it for the future, to use for drinking, washing, irrigating crops, and flushing toilets.

We will never be able to stop all the flooding that comes with climate change, but we can come up with innovative ways to manage it.

SEA-LEVEL RISE

REBECA SABNAM spent her childhood in Bangladesh, a country often described as "Ground Zero for Climate Change." Two-thirds of the low-lying South Asian country is just sixteen feet above sea level. As sea levels rise, so does the danger of flooding.

Climate change is causing Earth's oceans to rise. Average sea levels have gone up more than eight inches in the past century, and three of those inches occurred over the last two decades. Today, the sea is rising one inch every five years—the fastest we have ever seen.

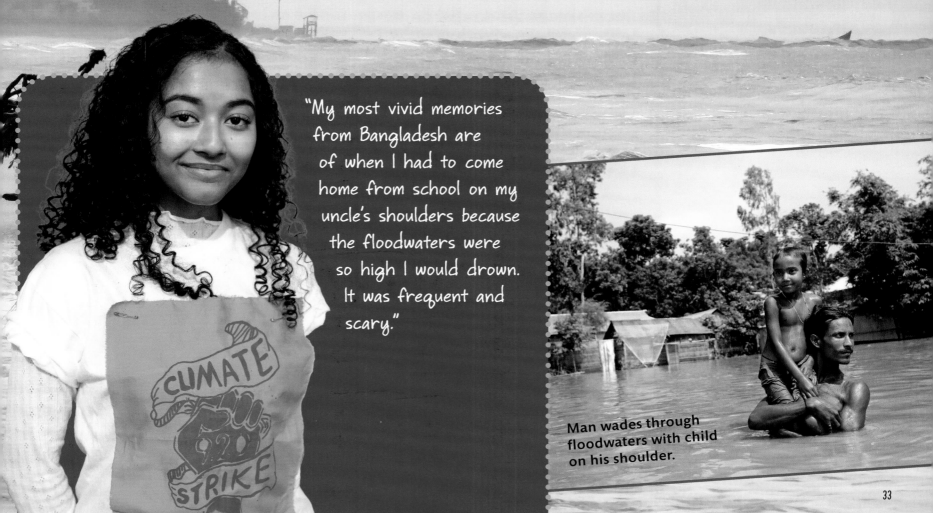

"My most vivid memories from Bangladesh are of when I had to come home from school on my uncle's shoulders because the floodwaters were so high I would drown. It was frequent and scary."

Man wades through floodwaters with child on his shoulder.

US COASTAL CITIES UNDER THREAT

In the United States, almost 40 percent of our population lives in cities and other densely populated communities on the coast. Rising sea levels make these people's homes vulnerable to storm damage and flooding.

MELTDOWN IN THE POLAR REGIONS

According to a recent NASA study, melting ice sheets and glaciers in the polar regions—Greenland and Antarctica—are causing one-third of all sea level rise.

The record heat wave that Europe experienced in 2019 caused Greenland's ice sheet to melt at a rate that scientists hadn't expected to see until 2070. Fifty-five billion tons of water melted from the Greenland ice sheet over five days in July and August 2019, which is enough to cover the state of Florida with almost five inches of water.

HOW DO WE KNOW THAT SEA LEVELS ARE RISING?

There are three ways that scientists measure sea levels.

✳ There are buoys anchored off all our coasts—they float on the water and measure how deep it is. By tracking high and low tides, scientists can calculate an average sea level every day.

✳ NASA uses satellites to measure the depth of the ocean. Even though they are orbiting nearly 1,250 miles above Earth, satellites record very accurate measurements, within one inch of the actual depth.

✳ We also track sea levels by putting large tubes called tidal gauges on the floor of the ocean and measuring how high the water rises up the tube.

Scientists are even more worried about Antarctica, where 90 percent of Earth's ice is located. Antarctica's ice is melting three times as fast as it was over a decade ago.

If all glaciers and ice sheets were to melt, NASA research reports that sea levels would rise by more than 195 feet. The threat of such extreme melting would be catastrophic. By the end of the century, rising seas would flood coastal cities around the world. New York City, New Orleans, Miami, Shanghai, and some entire island nations could become permanently flooded. The rising seawaters would also submerge vast stretches of surrounding lands and force 200 million people to evacuate flooded surroundings.

Iceberg "calves" break away from a melting glacier, falling into the sea.

GHOST FORESTS

Sea-level rise also creates "ghost forests." In the mid-Atlantic coast of the US, where sea levels are rising, the increasingly powerful Atlantic hurricanes push salt water inland. The salt water kills trees from the roots up, creating ghostly forests of bleached and blackened dead trees.

HOLDING BACK THE TIDE

Most coastal cities are planning to build seawalls that are designed to hold back storm surges that happen at high tide and cause flooding. These occur more frequently with the increasing number and severity of hurricanes. Some cities have already put barriers in place, like this one in London's Thames River.

GLOBAL CLIMATE STRIKE

From September 20–27, 2019, an estimated 7 million people participated in what is thought to be the world's largest protest demanding action on climate change. Young people and adults joined together to participate in over 2,500 events scheduled in 163 countries on all seven continents.

OCEAN ACIDIFICATION

In 2015, **LEVI DRAHEIM** was just eight years old, the youngest of the group of kids who sued the government. Climate change and the effect it is having on our oceans is top of mind for Levi and his family, who live on a barrier island, a narrow strip of land that separates the Florida coast from the Atlantic Ocean. The warming and acidification of the ocean is affecting sea life in their community, which relies on commercial fishing and tourism for their income and way of life.

Oceans absorb about one-third of all the carbon dioxide in our atmosphere. For millions of years, this exchange of carbon dioxide between ocean water and the atmosphere remained constant and helped to slow down global warming. But in the past 150 years, as humans have greatly increased the amount of carbon dioxide in the air by burning fossil fuels, the oceans have become 30 percent more acidic.

OCEAN ACIDIFICATION IS AN INVISIBLE THREAT

It happens when carbon dioxide from the atmosphere dissolves in seawater. That causes the oceans to become acidic. Acid eats calcium, which animals such as clams, mussels, crabs, and corals need to grow their exoskeletons and shells.

Ocean acidification affects all of us because as Earth's oceans reach the limit of the carbon dioxide they can absorb, global warming is accelerating. It also affects our food supply, as fish and shellfish are decreasing in numbers.

SAVING OUR OCEANS

The most effective way to limit ocean acidification is to act on climate change. We all need to look at our everyday choices and purchases and do everything we can to reduce our use of fossil fuels. If we can cut back significantly on greenhouse-gas emissions, we can improve the health of our ocean ecosystems.

TOP: Healthy coral on the Palmyra Atoll (ring-shaped coral reef) about a thousand miles south of Hawaii.

BOTTOM: Bleached and dying coral, caused by high water temperatures.

WILDLIFE DESTRUCTION

The main impact of global warming on wildlife is a **disruption in their habitats.** Animals have adapted over millions of years to changes in their surroundings and challenges to their habitats. But climate change is transforming habitats too quickly for animals to adapt.

POLAR BEARS

Certain kinds of animals are suffering more than others, and polar animals are particularly at risk.

It is not just polar animals that are affected. Orangutans that live in the rainforests of Indonesia are under threat as their habitats are cut down to make way for human development and more severe droughts lead to more bushfires. Sea turtles lay their eggs on nesting beaches, many of which are going to disappear because of rising sea levels.

A THREAT TO OUR FOOD CHAIN

Biodiversity (short for **biological diversity**) refers to the numbers of species of plants, animals, and microorganisms in all the places on Earth. It also refers to the variety of different habitats, such as forests, deserts, oceans, and coral reefs. Because all these systems interact with each other, species and ecosystem die-offs due to climate change can disrupt the balance of our diverse Earth, threatening the food chain for all animals as well as for us.

Endangered animals (clockwise from top left): Amur leopard, chimpanzee, sea turtle, dolphin, Bengal tiger, orca (killer whale), Bonobo.

HONEYBEE DESTRUCTION

Most alarming is the decline in the honeybee population in the United States. These insects, known as pollinators, travel from plant to plant, transferring pollen from one to another, fertilizing the plants. A German bee expert, Professor Jürgen Tautz, said: "Bees are vital to biodiversity. There are 130,000 plants for which bees are essential to pollination, from melons to pumpkins, raspberries, and all kind of fruit trees—as well as animal [food]—like clover. Bees are more important than poultry in terms of human nutrition."

The US Department of Agriculture says that one of every three bites of food that Americans consume is a product of bee pollination. Honeybees may be small, but they are mighty, pollinating about $15 billion worth of US food crops every year. Winter bee-colony collapse hit a high in 2018, with beekeepers losing 41 percent of their colonies, up 3 percent from the previous year.

IS EXTINCTION AVOIDABLE?

The current mass extinction is different from the events that caused dinosaur extinction sixty-five million years ago. Dinosaurs became extinct because of great natural changes in their environment. Today, animals are being threatened with mass extinction because of changes in their environment, but human actions are responsible for many of these changes. If people are the main cause of climate change, that means we can do something about it.

CONSERVATION:
GIVING BACK TO NATURE

One thing that many people are talking about is the idea of giving back to nature. Right now, about 3 percent of our oceans and about 15 percent of the land on Earth have been designated as protected areas—national parks are an example. We need to protect more areas on both land and sea.

Scientists and engineers are also focused on developing alternative ways to pollinate plants since bees are disappearing, including robots that roll up and down planted fields, using their tiny arms to pollinate the plants. We are also developing digital sensor networks that watch bee colonies and predict potential problems before they occur.

Researchers are working to create robotic bees
that fly and act much like real bees.

43

IT'S UP TO US!

You are a citizen of planet Earth. All of us live in natural environments around the world. Here are some things you can do:

APPRECIATE NATURE: Go to a park, or hike with your family and friends in nearby green spaces or natural areas located in a city, suburb, or the country and look at birds or view the colors of leaves changing from summer to autumn. You can visit a zoo or an aquarium to see some of the wild animals that share Earth with us. You can do so many things to learn to value nature and appreciate our planet.

TALK ABOUT IT: Discuss what you have learned about climate change with your friends and family. As you talk about it, try to link what's happening to a problem that people in your community are affected by and care about, like availability of clean water, wildfires, and air pollution.

BE A CONSERVATIONIST: Get your fellow students and teachers to work together to help your school reduce energy usage.

WORK WITH OTHERS: One person can do very little to affect climate change around the world, but you can vastly increase your impact if you join forces. That's what teenagers are doing with the School Strike for Climate, which started in Sweden but has spread around the world.

"Homo sapiens have not yet failed. Yes, we are failing, but there is still time to turn everything around. We can still fix this. We still have everything in our own hands." —Greta Thunberg

All the ideas in this book involve conservation. Keep reading, talking to people, and working to find out what new and innovative approaches we can try. Who is going to do that? It could be you. Perhaps the most important thing you can do is to grow up to be a responsible citizen. Continue to make good decisions and help find and carry out scientific and political solutions to the problem of climate change.

Remember:

YOU CAN MAKE A DIFFERENCE!

GLOSSARY

ACTIVIST A person who campaigns to bring about political or social change.

BIODIVERSITY The numbers and kinds of animals and plants in a place on Earth.

CARBON DIOXIDE A colorless, odorless gas produced by burning carbon and organic compounds and by breathing. It is naturally present in air (about 0.03 percent). Human activities are largely responsible for increase in atmospheric carbon dioxide beyond the naturally occurring levels.

CLIMATE CHANGE A change in global or regional normal climate patterns, attributed largely to the increasing levels of carbon dioxide in the air produced by the burning of fossil fuels.

DEFORESTATION The act of clearing a wide area of trees such as in the Amazon rainforest.

FOSSIL FUELS A natural fuel such as coal or oil, formed millions of years ago from the remains of living organisms.

GLOBAL WARMING A gradual and continual increase in the overall temperature of Earth's atmosphere scientifically attributed to the greenhouse effect caused by increased levels of carbon dioxide and other gases in the atmosphere.

GREENHOUSE EFFECT The trapping of the sun's warmth in Earth's atmosphere, caused by light from the sun easily passing through the upper atmosphere, then turning to heat rays at the surface and lower atmosphere.

GREENHOUSE GASES Gases that contribute to the greenhouse effect by absorbing heat radiation. Examples are carbon dioxide and chlorofluorocarbons.

HEAT WAVE A prolonged period of days or weeks of abnormally hot weather.

INDIGENOUS The first people who lived in a region; not later immigrants.

INDUSTRIAL AGE A period of changes in economic and social organization that occurred mainly in Europe and the United States. It began in the late 1800s and was caused mainly by the replacement of hand tools with power-driven machines, such as the power loom and the steam engine, and by the concentration of industry in large factories.

MASS EXTINCTION A widespread and rapid decrease in the kinds of animals or biodiversity on Earth.

OCEAN ACIDIFICATION The increase of carbon dioxide in the atmosphere, mostly from burning fossil fuels, is changing the acidity of our oceans. The oceans have become 30 percent more acidic over the last 150 years.

ORGANIC Matter from living things that contain carbon. It also refers to food including plants or meat that is grown or raised without chemicals or pesticides.

RENATURING Restoring something to its original state.

SEA LEVEL All of the oceans are one continuous body of water, so its surface seeks the same level throughout the world. However, winds, currents, river discharges, and different temperatures prevent the sea surface from being exactly level.

WATER VAPOR The gaseous phase of water. Water vapor is in the atmosphere along with droplets of liquid water. You can make water vapor by heating liquid water until it evaporates. The three states of water are vapor, liquid, and ice.

WEATHER The atmosphere at a particular time and place. It is described by its conditions such as temperature, humidity, wind velocity, precipitation, and barometric pressure.

WILDFIRE A large, very destructive fire that spreads quickly over forest or plains.

WILDLIFE EXTINCTION The process of a species or family of animals dying off and vanishing from Earth.

FOR FURTHER READING

- *What a Waste: Trash, Recycling, and Protecting our Planet* by Jess French (DK Children's, 2019)
- *What Is Climate Change?* by Gail Herman and WHO HQ, illustrated by John Hinderliter (Penguin Random House, 2018)
- *No One Is Too Small to Make a Difference* by Greta Thunberg (Penguin Random House, 2019)
- *The Boy Who Harnessed the Wind* (Picture Book Edition) by William Kamkwamba and Bryan Mealer, illustrated by Elizabeth Zunon (Penguin Random House, 2012)
- *Our House Is on Fire: Greta Thunberg's Call to Save the Planet* by Jeanette Winter (Simon & Schuster, 2019)
- Learn about youth movements online, like Fridays for Future. https://fridaysforfuture.org/

PHOTO AND ILLUSTRATION CREDITS

Dedication/p. 2: © NASA; p. 3: © Liz Nealon; p. 4: © Nur Photo / Getty Images; p. 6: © Nur Photo / Getty Images; p. 7: © Bryan Luna; p. 8: © Designua / Shutterstock; p. 9: © Nexus Media News; p. 10, foreground and background: © WENN Rights Ltd / Alamy Stock Photo; © titoOnz / Shutterstock; p. 11: © AccuWeather; p. 12, left to right: © US Department of Commerce/National Oceanic & Atmospheric Administration/NOAA Research; © Johan Swanepoel / Shutterstock; p. 13: © Sopotnicki / Shutterstock; p. 14, left to right, clockwise: © atsurkan / Shutterstock; © Suti Stock Photo / Shutterstock; © Lopolo / Shutterstock; © photo.ua / Shutterstock; © CharlesOstrand / Shutterstock; © PotapovAlexandr / Shutterstock; p. 15, foreground and background: © Avery Leigh White; © PhilipYb Studio / Shutterstock; p. 16, top to bottom: © Miko Bagus / Dreamstime; © US Department of Commerce/National Oceanic & Atmospheric Administration/Climate.gov; p. 17, left to right, clockwise: © Michael Jung / Shutterstock; © Richard Pratt / Shutterstock; © Mitchellsk / Shutterstock; p. 18, foreground and background: © Alexandria Villaseñor; © Christian Roberts-Olsen / Shutterstock; p. 19, top to bottom: © Joseph Sohm / Shutterstock; © IuliaIR / Shutterstock; p. 20: © Terray Sylvester / Reuters; p. 21: © Benjamin Rasmussen; p. 22, foreground and background: © Jaye Renold/If Not Us Then Who?; © Fletcher & Baylis / Science Source; p. 23: © Frontpage / Shutterstock; p. 24: © ricochet64 / Shutterstock; p. 25, top to bottom: © New Africa / Shutterstock; © AnastasiaKopa / Shutterstock; p. 26, top to bottom: © Brooke Anderson; © Kevin Lamarque / Reuters; p. 27: © UN Photo/Manuel Elias; p. 30: © Fotosr52 / Shutterstock; p. 31: © Ramit Plushnick-Masti; p. 32: © borchee / iStock Photo; p. 33, left to right, clockwise: © S B Stock / Shutterstock; © Mohammad Ponir Hossain / Reuters; © Avery Leigh White; p. 34: © NASA; p. 35: © Bernhard Staehli / Shutterstock; p. 36, top to bottom: © Judy Darby / Dreamstime; © Wei Huang / Shutterstock; p. 37, left to right, clockwise: © Joachim Zens / Shutterstock; © Holli / Shutterstock; © Ryan Rodrick Beiler / Shutterstock; © Holli / Shutterstock; © Ink Drop / Shutterstock; © Ryan Rodrick Beiler / Shutterstock; p. 38, foreground and background: © Robin Loznak / Zuma Wire; © Andrew J. Martinez / Science Source; p. 39, top to bottom: © Jim Maragos/US Fish and Wildlife Service; © Georgette Douwma / Science Source; p. 40: © B & C Alexander / ArcticPhoto / Science Source; p. 41, left to right, clockwise: © Anan Kaewkhammul / Shutterstock; © Ari Wid / Shutterstock; © Andrea Izzotti / Shutterstock; © dangdumrong / Shutterstock; © Monika Wieland Shields / Shutterstock; © Sergey Uryadnikov / Shutterstock; p. 42: © Andrew Mandemaker; p. 43: © National Science Foundation / NSF.gov; p. 44, left to right, clockwise: © Marina Poushkina / Shutterstock; © wavebreakmedia / Shutterstock; © Monkey Business Images / Shutterstock; © Jorik / Shutterstock; p. 45, top to bottom: © Elizaveta Galitckaia / Shutterstock; © Lev Radin / Shutterstock; p. 47: © Bryan R. Smith / Getty Images

INDEX

Bold type indicates illustrations.